Potcoin

The Cryptocurrency that will Revolutionize the Legal Marijuana Industry

that the author is not engaging in the rendering of legal, financial, medical or professional advice. The content within this book has been derived from various sources. Please consult a licensed professional before attempting any techniques outlined in this book.

By reading this document, the reader agrees that under no circumstances are is the author responsible for any losses, direct or indirect, which are incurred as a result of the use of information contained within this document, including, but not limited to, —errors, omissions, or inaccuracies.

Table of Contents

Benefits of Using PotCoin

How to Use PotCoin

Consumers

Investors

Advantages of using PotCoin as a Payment Method

1. Safe
2. Fast
3. Low Transactions Fees
4. Global
5. Map Listing

Chapter Four: How to Buy, Sale and Store PotCoin

How to Buy PotCoins

Steps in Buying PotCoins

1. Create an Account with PotWallet.com
2. Create an Account with an Exchange that Allows you to Trade POT
3. Deposit Funds into Your Exchange Account
4. Buy POT

Introduction

With the legalization of recreational marijuana in 9 states, including the District of Columbia and medical marijuana in 29 states, retail sales in marijuana in more than 20 states reached a high of $6.5 billion with the number expected to rise to $30 billion by 2021 and $50 billion by 2026.

With this massive figure, a cryptocurrency innovation came into place to deal exclusively with the steadily growing marijuana business. PotCoin is a banking solution that brings the marijuana business and consumers together on a decentralized platform, allowing participants around the globe to make secure transactions.

PotCoin is still a developing

cryptocurrency and to start; a user creates a digital wallet which generates a unique public address to receive the coins and a private key to access the funds.

The PotCoins are transferred to a user's wallet and can be used anonymously to buy and sell cannabis products on a global scale. This book will delve deeper into PotCoins, analyzing its history, technology, how to buy, sell and store it. We will also look at whether PotCoin has a future in the already-crowded cryptocurrency market.

Chapter One: Introduction To Potcoin

PotCoin (code: POT) is a peer-to-peer cryptocurrency that was innovated with the aim of becoming the standard way of payment for the legalized industry of cannabis.

The developers, based in Canada, created PotCoin with the mission "to empower, secure and facilitate the legal cannabis community's transactions by creating a unique cryptocurrency."

This digital currency allows marijuana consumers to buy and sell cannabis products anonymously and provides a banking solution that brings the marijuana businesses and consumers together on a decentralized peer-to-peer platform, allowing participants all around the world to make transactions. PotCoin was released under the MIT/X11 license which was technically like Bitcoin under an open source infrastructure, which means that improvements and changes to its code can be made by the currency's administrators or external supporters. In 2017, PotCoin moved away from the mining system like Bitcoin to a proof of stake (PoS) system whereby participants earn 5% to 7% interest on their PotCoin holdings and transaction history.

As with most cryptocurrencies or any medium of exchange, the value of PotCoin is directly proportional to its demand, or how much transactions are carried out in a single day. As of today, PotCoin does not have a large community behind it as compared to Bitcoin or Ethereum, thereby making it volatile in the cryptocurrency market. However, much publicity, such as sponsoring former basketball player Dennis Rodman's trip to North Korea on June 13, 2017, increased PotCoin's value by 64.35% from $0.104 to $0.1723 on the same day. PotCoin has received minimal media coverage from media agencies such as Fox Business, Vice, and TechCrunch.

History of Cryptocurrency

The history of cryptocurrency is taught to have begun in 1983 with the American cryptographer David Chaum who invented anonymous cryptographic money called ecash. In 1995, he developed it further to what he called Digicash, which was an early form of cryptographic electronic payments which required user software to withdraw notes from a bank and designate certain encrypted keys before it could be sent to a receiver. This being the reason, the currency could not be traced by the issuing party, government or a third entity.

A paper was published in 1966 by NSA entitled "How to Make a Mint: The Cryptography of Anonymous Electronic Cash." The article described a Cryptocurrency system, first publishing it in an MIT mailing list and The American Law Review (Vol.46, Issue 4).

A description of "b-money," an anonymous, distributed electronic cash system was published in 1998 by Wei Dai. Shortly afterward, BitGold was created by Nick Szabo, which is now referred to as the father of cryptocurrencies. BitGold was an electronic currency system which required users to complete a proof of work function with solutions being cryptographically put together and published.

However, the first decentralized cryptocurrency was Bitcoin which was created by an anonymous developer, Satoshi Nakamoto in 2009, introduced through a paper called "Bitcoin - A Peer to Peer Electronic Cash System" posted on a mailing list discussion on cryptography. Nakamoto used an SHA-256 cryptographic hash function as its proof-of-work scheme. The second known cryptocurrency, Namecoin, was created in April 2011 in an attempt at forming a decentralized DNS, which would make internet censorship very difficult. In October 2011, Litecoin was released being the first successful cryptocurrency to use scrypt as its hash function instead of SHA-256.

Cryptocurrency has undergone major

changes leading to the establishment of some coins including Ethereum, Bitcoin, and Litecoin which are the major cryptocurrencies holding the highest values in the crypto world. Crypto coins can be used to buy products and services with the first Bitcoin ATM launched by Jordan Kelley in the United States in February 2014. By September 2017, more than 1500 Bitcoin ATMs have been installed around the world with an average fee of 9.05% with an average of 3 Bitcoin ATMs being installed per day as of September 2017.

History of PotCoin

The history of PotCoin is rather short. Three entrepreneurs from Montreal, Canada who identified themselves as

Hasoshi, Mr. Jones and Smokemon 514, developed and released PotCoin on January 21, 2014, via GitHub with the aim of creating a common cryptocurrency for the ever-growing legal cannabis industries across the globe. The objective, according to the PotCoin whitepaper, was to develop a blockchain coin dealing in cannabis which will be focused on marketing marijuana, creating a business partnership and establishing a merchant relationship in the legal marijuana industry globally. The developers were also tasked to explore the possibilities of smart contracts, and how it can benefit consumers and merchants in the legal cannabis industry.

By the cannabis culture, PotCoin was launched at exactly 4:20 p.m. and, within a week of its launch on GitHub, PotCoin garnered sufficient interest to warrant multiple mining tools. In the next month after its successful launch, PotCoin was added to the clearing/exchange Cryptocrush, such that the trading between Bitcoin and PotCoin was now possible. With a large growing community and an expanding team of developers as of February and March 2014, PotCoin started to gain mainstream media attention that some cannabis merchant announced that they would start accepting PotCoin as a means of payment for marijuana supply. For instance, Chronic Star Medical, a company dealing in the supply of cannabis edibles, such as drinks and cakes, was the first company to announce PotCoin as a means of payment on February 17, 2014. As of March, PotCoin announced that they had been added to the three cryptocurrency exchanges accounting to the largest trade volumes to date. Additionally, the development team announced that they had secured their first seed-round investment, indicating a rise in investors and the exchange rate.

PotCoin developers for the very first time revealed their identities to the public in the New York Cryptocurrency convention where co-founder Joel Yaffe and developer Nick Iversen gave a public talk on PotCoin at the conference which was attended by thousands of cryptocurrency developers, investors, and enthusiasts. The team also announced that they would be present in Denver for the 420-counterculture holiday in which PotCoin witnessed a dramatic rise in price, taking its market capitalization by over 1 million USD. This was attributed to an increase in investment due to the excitement around the April 20th counterculture holiday.

However, with an increase in investors due to increased investment speculation, PotCoin systems experienced a significant crash on 20th April 2014 creating a sense of doubt among the investors. This attributed to a decrease in PotCoin market capitalization which sunk to a low of $244,000 on May 22, 2014.

Nevertheless, PotCoin made a recovery and overturned the downtrend in price with the installation of PotCoin ATMs in two River Rock Wellness dispensaries in Colorado on May 27, 2014. On July 1, 2014, PotCoin realized an all-time capitalization of $1,860,000 attributed to a growing legal marijuana industry coupled with an increase of investors.

PotCoin yet again experienced difficult economic times in early 2015 after internal conflicts erupted leading to the development team breaking apart for various reasons not specified. However, the Marijuana community tried to rescue PotCoin from death making it a community-driven coin. Notably, PotLabs, who were a group of PotCoin enthusiasts, took charge of the community-driven currency and developed it further, pushing to keeping it alive, valuable and in circulation.

So much did the PotLabs develop the coin, that on August 23, 2015, they released an updated version of PotCoin which was highly anticipated, with an updated version which was based on POSV algorithm, the PotCoin system's speed lead to an increase in investors. Before this, there were some issues with the network stability causing many cryptocurrency exchanges to freeze their PotCoin wallets as they were unsure of the coin's future. However, with the normalization of the network after a few weeks, the exchange wallets were unfrozen allowing normal transactions to resume.

To date, PotCoin has continued with massive advertisement and promotions, notably being sponsoring Ex-NBA star Dennis Rodman's to the North Korea summit on June 13, 2018, being the fifth trip in which he was sponsored. Dennis Rodman's trip to Singapore wearing a PotCoin branded T-shirt for the North Korea summit, in which Kim and Trump historically met, made waves in the mainstream media with CNN featuring him as a guest, resulting in many online responses. At the summit, Rodman personally met with Trump and Kim leading to high publicity which, in turn, promoted PotCoin so much that it grew in value by 23% in one day, trading at

$0.0965. As of June 6, 2018, PotCoin was exchanging at $0.075191 and 0.00001147 BTC per 1 PotCoin. The circulating supply was 220,127,979 POT, with a market cap of USD 16,552,687 equal to 2,524 BTC. Does PotCoin have a clear future? Let's continue with our discussion.

Chapter Two: The Technology Behind PotCoin

Blockchain 2.0 Technology

Just like Bitcoin or most cryptocurrencies, PotCoin is based on the blockchain 2.0 technology. A blockchain refers to a continuously growing list of records, known as blocks, which are linked and secured using cryptography. Each block is made up of a cryptographic hash of the previous block, a timestamp, and transaction data. A transaction between two parties is recorded in an openly distributed ledger. The recorded transactions are resistant to modification of data, are permanent and can be verified. A blockchain is managed by a peer-to-peer network, which collectively adheres to an inter-node protocol for communication and the validation of new blocks. In a blockchain, information cannot be altered retroactively without alteration of all subsequent blocks, which requires a consensus of the network majority, thereby safeguarding the whole system. The blockchain technology behind PotCoin has achieved a decentralized consensus, thus making it suitable for peer-peer transactions. Before the release of PotCoin, 55 blocks were mined for checkpoints and analysis. The initial blockchain reward was set at 420 PotCoins but, on June 1, 2014, the block reward was cut in half and currently stands at 210 PotCoins.

PotCoin operates under the MIT/X11 license which was technically identical to Litecoin until August 23, 2015, when PotCoin changed to Proof-of-Stake velocity

(PoSV) similar to Redcoin. Initially, PotCoin was virtually similar to Litecoin. It was a marijuana-themed version of Litecoin, built like a fork of Litecoin-QT. The main improvements over Litecoin were a shorter block generation time, a quicker halving schedule, and an increased maximum number of coins. The MIT license refers to a permissive license originating at the Massachusetts Institute of Technology (MIT). The license being permissive puts only insufficient restrictions on reuse and has, therefore, excellent license compatibility. However, MIT license permits the use of the proprietary software if all copies of the approved software encompass MIT license terms coupled with the copyright notice

The technology behind PotCoin is also based on the Ethereum Decentralized Network as well as other platforms, such as Counterparty and Rootstock. The Ethereum virtual machine allows the creation of Decentralized Autonomous Organization (DOA) which allows for community voting, a decentralized development, the creation and implementation of blockchain 2.o technology and smart contracts. Additionally, PotCoin is not managed by any central authority and provides a decentralized solution for the transfer of value.

Ethereum Network

PotCoin is anchored into the Ethereum network which is seen as a Virtual Super Computer and investors have to connect to this network to facilitate a PotCoin transaction. The Ethereum network is the basis for decentralized consensus and is the peer-to-peer network of participating nodes which maintain and secure the blockchain. The Ethereum network is based on a dashboard of live statistics referred to as EthStats.net. This dashboard displays essential information including the current blockchain, in this case, PotCoin, hash difficulty, gas price and gas spending. The nodes shown on the network pages are only a selection of actual nodes and networks. Anyone is allowed to add information regarding a node to the EthStats dashboard through EtherNodes.com which displays the current and historical data on the node count and other information on both the Ethereum mainnet and Morden testnet.

PotCoin developers viewed the Ethereum network as the most suitable blockchain network offering a greater possibility for the growth of PotCoin. The Ethereum platform offers a significant advantage to the growth, scaling, and development of the cannabis coin; one of them being securing the coin from hackers or unauthorized transactions. Additionally, the Ethereum network allows PotCoin to leverage the security and development benefits of the cannabis industry to its ever-growing users and merchants; however, there are some significant risks associated with PotCoin being anchored on the Ethereum network. For instance, the Ethereum network is immature and still developing daily and therefore can experience failure in its entire system or be exposed to security threats. PotCoin specification can be listed as below:

- PoW algorithm: Scrypt, NOW POSv

- The total amount of coins: 420 million POTs

- Coins Per Block: 420

- Block Target: 40 seconds

- Retarget: 107 blocks

Mining PotCoin

With the inception of PotCoin, mining was through a computer's hardware including CPUs, GPUs, and, more commonly, scrypt ASICs. However, on August 23, 2014, PotCoin changed mining algorithms from Scrypt to POSV, proof of stake algorithm created to reward not only the ownership of coins but also the transactions. This meant that the coin would no longer be supported by mining, but instead, the coin will be generated by a network of QT wallets earning interest. Additionally, new coins will be generated and paid out to wallets for supporting the network.

Apart from the technology behind PotCoin, the company also offers a range of cannabis products which can be obtained at the PotCoin shop as discussed below.

PotCoin Shop

The PotCoin shop offers a variety of cannabis products, such as the crypto juice, which is sold directly to the public with the use of PotCoin or fiat as a medium of exchange. All purchases made using the PotCoin receives a 33% to 50% discount. The crypto juice is an energy-powered cannabis energy drink. The drink has a natural test to keep one energized throughout the day and can be mixed with a wide variety of juices and flavors.

PotCoin Seed Bank

The PotCoin Seed Bank offers a simple solution to turn your PotCoins into real cannabis seeds. To convert your PotCoin into real cannabis seeds, one can simply join The PotCoin Seed Bank, stake their PotCoins which will later generate "SEED" tokens based on the number of seeds available in the bank. Each SEED token is redeemable for one cannabis seed from the PotCoin Seed Bank which operates on a daily basis, and the cannabis seed will be shipped free of charge. This is after PotCoin partnered with a leading cannabis seed producer to offer an opportunity to the PotCoin community and cannabis enthusiasts. Based on this system, PotCoin holders will be eligible for generated seeds based on their holdings; therefore, the more PotCoins one can stake, the more seeds they will earn.

Anyone can join the PotCoin seed bank, but they will need to be participating geolocations listed to be able to receive the cannabis seed. Once a PotCoin address is accepted into the Seed Bank, the coins must be staking at that address for about 30 days to generate seeds. The amount of seeds generated depends on the total number of Seed Bank members as per the table below:

Example the Bank	Seed Bank PotCoins	Seeds Generated	Seed in per 100k PotCoins
Situation-1-	1,000,000	250	25 seeds
Situation-2-	2,500,000	250	10 seeds
Situation-3-	25,000,000	10,000	1 seed

PotCoin Reward Program

PotCoin came up with a reward program aimed at promoting the coin. In the program, PotCoin allows merchants and business to reward their current clients to use and familiarize themselves with PotCoin by giving free PotCoins with every purchase of a cannabis product made. To achieve this program, the PotCoin development team partnered with existing businesses in targeted geographic locations which are best suited for PotCoin. The merchants then receive free PotCoins and redeemable coupons codes which can be distributed to existing customers. The PotCoins given to the developmental team is obtained from the Development Fund which is contributed to by the community. The table below shows a sample of the reward program:

Example	# of PotCoins	Customers	PotCoins per Customer
Level-1-	250,000	2,500	100POT
Level-2-	500,000	5,000	100POT
Level-3-	1,000,000	10,000	100POT

PotCoin is still developing, and the core development team continues to add certain modifications aimed at improving the project. The developing team continue to add to the project and is actively seeking out partnerships in Canada where it was incepted internationally.

Chapter Three: Uses of PotCoin

Provides a Decentralized Banking Infrastructure for the Legal Marijuana Industry

With many legal reforms fast approaching in the cannabis industry, from California allowing recreational use on January 1, 2017, and Canada legalizing it federally, it is certain that PotCoin has some uses. The main use, however, is to empower, facilitate and secure transactions in the legal marijuana space. PotCoin is the first association between a cryptocurrency and licensed cannabis producers who can be vendors and suppliers. The coin network allows cannabis enthusiasts to interact, transact, communicate and grow together. PotCoin offers a solution to the marijuana industry which has continued to be hugely neglected by most governments despite its proven medical and recreational value. The market worth of the legal cannabis industry is estimated to value 24.1 billion U.S. dollars, of which 13.2 billion dollars is for medical use. This figure is expected to increase as more states are expected to pass laws legalizing the medical and recreational use of marijuana, with most countries in Europe and around the globe also to follow.

PotCoin offers a standard, secure and decentralized means of exchange in the cannabis space. Since marijuana remains illegal under federal laws, pot shops and related enterprises are unable to use the traditional financial services like credit card providers and banks to take care of business. Additionally, the financial institutions are afraid of being prosecuted for money laundering or similar charges by the Federal Bureau. None of the bills filled to address the marijuana banking challenge have ever been implemented. Thus, many of investors in the cannabis space are operating on a cash-only basis, which is not only inconvenient, but it can also be dangerous as it can lead to money laundering charges. Also, cannabis businesses lack a non-physical way of completing secure transactions attributed to its notion of illegality from the general public who may not want to be associated with it.

PotCoin just offers the perfect solution to the problem of marijuana banking and also offers a secure transaction, eliminating the use of a cash-only basis which is inconvenient and exposed to fraud. PotCoin provides a banking solution that brings marijuana business and consumers together on a decentralized peer-to-peer platform, which allows participants all over the world to make safe transactions. To start, a user creates a digital wallet which creates a unique public address to accept the coins and a private key to see the funds. PotCoins transferred over to a user's wallet can be used anonymously to purchase and sell cannabis products on a world-wide scale.

However, PotCoin is limited regarding what one can purchase with it. Even though the coin's development team claim that transactions with the coin are anonymized, there is lack of anonymization for the cryptocurrency. This is because the currency's specific use in the legalized cannabis space limits its user base to only marijuana consumers and dealers; therefore, if an investigating body or the government decides to round up marijuana dispensaries, they only need to dig out transactions made under the banner of PotCoins. The data will be obtained from the PotCoin blockchain which would be easier to decrypt as the volume wouldn't be as high as other transactions carried out with other cryptocurrencies such as Bitcoin and Ethereum.

Nevertheless, PotCoin provides the underserved legal marijuana with a decentralized banking infrastructure and payment solution by eliminating the over-reliance of cash as a means of exchange which could be dangerous and associated with money laundering charges. Through the use of PotCoin, the players in the 100 billion legal marijuana industries will realize cost savings, scalability, and unparalleled enterprise security.

Benefits of Using PotCoin

PotCoin offers some benefits to investors and consumers in the legal multibillion marijuana space. Some of the benefits include:

• PotCoin can instantly be transferred over the internet or a system of networks allowing to efficiently and easily complete transactions.

• Since PotCoin is based on the blockchain technology which offers significant benefits to cryptocurrencies, including quality assurance, smart contracts, elimination of errors in accounting and peer-to-peer global transactions, it will significantly improve the legal marijuana sector by building a secure, global and transparent business platform.

• PotCoin is transferred directly from person to person via the net, without going through a bank or online money transfers, clearing agents who usually charge additional fees to complete the transactions. Therefore, transactions with PotCoin are inexpensive as low fees include reduced processing and security costs, leading to increased growth and scalability. Additionally, PotCoin eliminated chargebacks for merchants, further reducing additional expenses for marijuana sellers.

How to Use PotCoin?

As a currency, PotCoin can be used by both consumers and businesses.

Consumers

Consumers can easily use PotCoin by following the steps below:

1. First, pick a wallet, notably being a PotWallet, although there are other wallets such as Poloniex, Bittrex, PotCoinTrade, and CoinPayments. It is free to create a wallet and can be set up in minutes. Most wallets are available for Android and desktop operating systems. At the time of writing, PotWallet was available for Android and Windows, while it was still being developed for Mac OS X, and Linux.

2. The second step is to obtain PotCoin through an exchange, or by selling goods or services online.

3.   You can spend, stake or store PotCoin. PotCoin cannot be mined; rather, it uses a proof of stake mechanism, which means you can earn interest by holding onto PotCoin in your wallet.

4.   By staking your PotCoins, you can earn an annual interest rate of about 5%. This is because your computer becomes an active node on the network. You receive a reward in the form of additional PotCoins.

Investors

Legal cannabis-based businesses can accept PotCoins through the PotCoin merchant gateway. The merchant gateway is targeted towards licensed producers, dispensaries, or anyone who sells cannabis-related products. Additionally, any merchant can accept PotCoins regardless of their industry. Marijuana businesses who have a physical location can sign up on PotWallet to accept PotCoin as a means of exchange at their business premises using PotWallet's merchant system. The coin is paid directly into their PotWallet account. Similarly, online businesses (e-commerce) can sign up to CoinPayments to accept PotCoin on their e-commerce store using CoinPayment's easy plugins for all major platforms including Magento and WordPress.

PotCoin has a number of advantages including Some of the perks of PotCoin, in addition to using a frictionless currency with low fees, include a map listing on PotCoins.com's map. According to its white paper, PotCoin will eventually feature a map of all businesses that accept PotCoin.

Advantages of using PotCoin as a Payment Method

There is significant value in accepting PotCoin as a payment method as discussed below:

1.   Safe

Many businesses in the marijuana industry accept cash only as a method of payment, thereby putting their income at a liability of being stolen. Cash only also put customers at risk by making them carry hard cash which puts them at risk of losing it through various frauds. PotCoin is digital and decentralized, so funds can be transferred seamlessly from the customer's digital wallet to the business's wallet keeping both the investor and customer safe.

2. Fast

Unlike credit card payments that take a few days to settle into your accounts, PotCoin payments are accepted in seconds and settled immediately. This ensures that users can use their free "cash" flow when you need it, not when they feel you should have it.

3. Low Transactions Fees

As mentioned above, banks and online fund transfers charge hefty fees to process transactions adding additional expenses. Compared to the 6-10% you're being charged to accept credit cards, accepting PotCoin will cost you less than 1%. Additionally, there are no chargebacks since PotCoin is a digital currency or a cryptocurrency. Any confirmed transactions made using PotCoin is protected by the network and therefore can't be reversed. This effectively protects the merchants from fraud and saves a business's revenue by about 2% which is usually incurred using standard payment options.

4.   Global

Like any cryptocurrency, PotCoin is globally accepted and therefore a business does not have to deal with exchange rates which could be costly and could incur a loss. Additionally, accepting PotCoin as a payment method can make a business expand into new territories without needing accounts for each currency. With this, gaining new audience or customers is pretty easy without having to carry extensive advertisement or any other product promotion campaigns.

5.   Map Listing

Accepting PotCoin as a means of payment can get a business listed on the interactive map of accepted PotCoin merchants, which will be visible not only through PotCoin.com, but several other PotCoin-related sites that will provide the information to their audience. Through a map list, a marijuana business will be placed on a global scale, significantly increasing sales.

PotCoin will certainly revolutionize the marijuana industry by eliminating the need for cash and making purchasing marijuana safer for customers and sellers alike.

Chapter Four: How to Buy, Sale and Store PotCoin

PotCoins can be bought, sold and stored using a cryptocurrency wallet, which is equivalent to a normal wallet. A private key can be used to generate a personal address in addition to redeeming contents of a wallet. The public key is a unique identifier of 26-34 alphanumeric characters and is referred to as one's address. PotCoin wallets, with the most common one being PotWallet, can be a desktop software, a mobile application or can be hosted on the internet.

PotCoin-Qt, which is an open-source PotCoin client, is used to accomplish timely payments in addition to being a server utility for business.

PotCoin wallets are hosted by numerous

online services, allowing users to create a personal wallet account, store PotCoins and remotely access their coins through an app on their smartphone. The wallets are hosted on by the services on their servers enabling users to deposit and withdraw a balance from their website.

PotCoins can also be stored in cold storage. This is possible through the use of physical items to store a private key, a common example being paper wallets. To redeem and send PotCoins held in cold storage, the private key must be imported and stored in an electronic wallet. In this context, we are going to use PotWallet to discuss how to buy, sell and store PotCoins.

How to Buy PotCoins

PotCoins can be bought on its very own Potwallet.com peer-to-peer platform and some popular cryptocurrency exchanges including Changeling, Cryptopia, and many other wallets. PotWallet accepts credit cards and debit cards as means of deposit, with the flat currencies being USD and EUR. Below are steps followed in purchasing PotCoins using PotWallet:

Steps in Buying PotCoins

1.   Create an Account with PotWallet.com

Setting up your own personal account with PotWallet is simple. You just have to enter your personal details and register the account with a valid email address. A two-factor authentication system will then inquire your password to keep your personal details secure. Once on the platform, it is trouble-free to find PotCoin traders. You just have to add your location, and then search through the traders offering PotCoin in your location, state or country. Users can purchase PotCoin with PayPal, Bitcoin or international wire (Swift).

It is important to note that traders on the platform are independent of PotWallet and each has their own criteria for trading. For instance, a seller might request a copy of your personal ID before handing over their coins; therefore, you should make sure you analyze their exchange criteria before embarking on a trade.

However, if there is an issue with a trade, PotWallet will flick the status to" Dispute" and will retain the coins in escrow until the financial aspects of the trade are finalized/resolved, or both parties are in agreement with the terms of the trade.

2. Create an Account with an Exchange that Allows you to Trade POT

This step can serve as an alternative to step one. You may choose to find an exchange that offers PotCoin currency pairing; however, since PotCoin is still a developing coin with a minimal number of followers, it is not widely listed on cryptocurrency exchanges. Therefore, it may be hectic to find a legit exchange. Nevertheless, once you have found an exchange, you can sign up for an account by providing your name, email address, and password.

Under Know Your Customer (KYC) laws, some exchanges will need you to supply proof of ID to verify your account before you can start trading. It is recommended that you use two-factor authentication as it will provide an extra level of security for your accounts.

3.   Deposit Funds into Your Exchange Account

The next step is to deposit funds into your exchange account. As of the time of writing, only the Changelly wallet directly exchanged US dollars (USD) to PotCoin; therefore, if you choose a different exchange, you'll need to own or buy a cryptocurrency that is available in a currency pairing with PotCoin, such as Bitcoin (BTC), and then exchange it for POT.

To deposit BTC into your PotWallet trading account, look for the relevant "Deposit" link from the account homepage and follow the prompts to create your wallet address and transfer your funds.

4.   Buy POT

Finally, once the funds have been successfully deposited, search for the currency pair you want to trade. After you have found POT/BTC, you can then click on "Buy POT" to see the market price.

It is then a matter of choosing a limit or market order and entering the amount of BTC that you would like to spend or the amount of POT you are hoping to buy. Some exchanges will instantly place your order as soon as you click "Buy" button, while others will give you a chance to check the details of your transaction before finalizing your purchase.

How to Sell PotCoins

The process of selling PotCoins is much the same as that of buying it. However, in selling PotCoins, it is important to consider that not all currency pairing will be available on any given exchange, so that it may not necessarily be possible to exchange POT for your desired currency right. Here are the steps in selling PotCoins with PotWallet:

Steps in Selling PotCoins (With Advertisement)

1. Register on PotWallet as indicated in the previous sections.

2. There are two selling sections: Sell PotCoins online with your online banking or cash deposits and sell

PotCoins with cash and meet the local seller face-to-face. Thus, select the most appropriate method.

3. Go to Post an AD section.

4. Fill the form with your currency, location, payment method and your price margins you wish to sell and then post your ad.

5. Your PotCoin trade advertisement should now be published.

6. When a trade is initiated, you can contact the buyer through the trade messaging interface and arrange your payment.

7. Once payment is received, click on" Release Escrow."

Steps in Selling PotCoins (Without Advertisement)

1. Register on PotWallet.

2. There are two selling options: Sell PotCoins online with your online banking or cash deposits or sell PotCoins with cash and meet the local seller face-to-face.

3. Go to the Sell PotCoins sections.

4. Fill in the form with your location, payment method and the amount you wish to sell.

5. Select from a list of PotCoin traders available in your region. You can filter these by your desired payment method.

6. From the trader's list, choose a trader with a good reputation score

displayed after the username.

7. Press sell and send a trade request with a message to the trader. For online trades, the buyer will provide his payment details once the trade is opened. For cash trades, you should agree with the meeting time and place to conduct the trade.

8. Contact the buyer directly through the trade messaging interface and arrange payment.

9. Once payment is received, click on "Release Escrow."

How to Store PotCoins

You can only store PotCoins on a PotWallet which can be synced on your desktop computer. Storing PotCoins come with an advantage as your PotCoin holdings can generate an annual interest rate of about 5%. By storing and staking your PotCoins, your computer becomes an active node on the network, which, in turn, rewards you with additional PotCoins. Desktop wallets are the best for storing your coins as they make up the foundation of the PotCoin network. Additionally, they allow you to exercise complete control of your money at the highest level of security and earn interest with your PotCoins.

It is important to sync your desktop wallet before storing your precious coins in it. This is done in two separate different ways depending on the comfortability in performing the steps and the time consumed. One way requires changing multiple files, the other only changing one file. Either way, it's up to you to choose, following the steps below for the Windows operating system.

Method 1: Replacing Three files: Zipped Blocks, Chainstate, and Peers.dat file

1. Download the Windows desktop wallet.

2. Launch the wallet, and this will provide you with an incorrect date which is many weeks behind (this is normal). Usually, this takes several days to sync, but you'll be greatly accelerating this process by manually adding the blockchain rather than letting it sync on its own.

3. Shut down your computer before your next step.

4. Open the folder located in the following path: C: users>admin (or your   user

account)>AppData>Roaming>PotCoin. If you cannot find the AppData folder, it's because it's hidden, so you need to show hidden files and folders.

5. Download the zip folder containing Block folder, Chainstate folder, and Peers.dat file.

6. From the zipped file, you have just downloaded, extract its contents.

7. Replace any files that currently exist in the PotCoin folder with those from the zipped file.

8. Launch the PotCoin desktop wallet and wait for the splash screen which will show for at least 10 minutes. You should be patient while the blockchain fully loads.

9. Once the splash screen disappears,

and the client has fully loaded, you will notice that blockchain is a couple of days behind, instead of 190+ weeks. It should now only take less than one hour to complete the sync, instead of several days.

Method 2: Replacing only one file-bootstrap.dat

1. Download the Windows desktop wallet.

2. Start the wallet, and you see it's a little bit backward in time. Normally, it takes a few days to sync or rather, you could integrate the wallet to a blockchain to launch it effectively over a short time.

3. Download bootstrap.dat file here which contains the blockchain files.

4. Make sure the desktop client is closed.

5. Move the bootstrap.dat file to C:users.admin (or your user account)>AppData>Roaming>PotCoin.

6. Launch PotCoin desktop wallet and wait! The splash screen will show for a couple of minutes depending on the computer's processing power.

7. Once the splash screen disappears, and the client has fully loaded, you will notice that the client is "importing blocks from disk." This process can take a number of days to complete.

8. Once the wallet is fully synced "(out of sync)," the error message will disappear, and your wallet will be

ready for use.

Benefits of Storing PotCoins

There are benefits of storing your PotCoins which include:

1. Earn Interest

When you store PotCoins, the network rewards you with an annual interest of approximately 5% per annum which can be very beneficial as a piece of investment.

2. Support the Network

By storing your PotCoins, your computer becomes an active node on the network, processing and relaying transactions. It not only earns you additional PotCoins, but it's also the backbone of the PotCoin network and security.

3. Control your Funds

Storing your PotCoin requires the coin to be held in your desktop QT wallet. This gives you complete authority over your coin and encourages you to take control of your finances rather than leaving the decisions of third parties, including banks, to take control of your hard-earned coins.

Chapter Five: Legalization of Marijuana and the Potential Rise of PotCoin

Legalization of Marijuana

One of the most vexing, fiercely debated issues in the world today is whether to legalize marijuana for both medical and recreational use or to illegalize it all together. While the medical value of marijuana, such as relieving pain from multiple sclerosis, relieving insomnia, anxiety, spasticity and treating potentially life-threatening conditions such as epilepsy is well known, proven and documented, there is still little efforts put forward to legalize it or rather marijuana legalization faces huge resistance from policymakers, church leaders and many people in society. Nevertheless, many civilian human right bodies are for the legalization of marijuana. In the United States, for instance, the Drug Policy Alliance believes marijuana should be eliminated from the criminal justice system and regulated similarly like alcohol and tobacco.

Despite President Donald Trump promising not to interfere with people who use medical marijuana during the presidential campaigns, contrary to his promise, his administration is currently threatening to reverse this policy. With 29 states legalizing the medical use of marijuana and nine states allowing recreational use excluding, notably, Washington, DC, marijuana is still seen as illegal from the federal government's perspective. About 85% of Americans support law reforms legalizing marijuana in addition to carrying out certain initiatives aimed at legalizing marijuana. For instance, the Drug Policy Alliance (DPA) is working on marijuana legalization campaigns in New Jersey, New Mexico, and New York. DPA has continually pushed for the legalization of marijuana through a constitutional amendment that will allow its sale, possession and both recreational and medical use.

Around the globe, the legality of marijuana varies from country to country. Marijuana use and possession is illegal in most countries and has been since the agreement about hashish, in the International Opium Convention (1925) and the beginning of widespread cannabis prohibition in the late 1930s. However, the possession of cannabis in small quantities has been decriminalized in many countries and sub-national entities in several parts of the world. Uruguay became the first country to legalize selling, cultivation, and the distribution of cannabis on 10 December 2013. In 1976, the Netherlands passed the Opium Law which enables marijuana consumers to buy marijuana in legal outlets referred to as "coffee shops" if certain rules are followed, but large-scale production and trade remain illegal.

The medical use of Marijuana is legal in some countries, including Canada, the Czech Republic, and Israel. Medical marijuana in the United States is legal in 29 states as of December 2016, with efforts being made by civilian bodies in fighting for its legalization. However, in the United States, the Obama Administration refrained from prosecuting users and dealers operating in compliance with the state's cannabis laws.

In most countries, the laws governing the illegality of marijuana varies greatly. Some countries have laws which are not vigorously prosecuted as other countries, but, apart from the countries which offer access to medical marijuana, most countries have penalties ranging from lenient to serve. Some violations of the marijuana laws are taken more seriously in some countries than others regarding cultivation, use, possession or the transfer of cannabis for recreational use. A few jurisdictions have smaller penalties for possession of smaller quantities of marijuana, making it punishable by confiscation and also a fine, rather than imprisonment. Contrary to some countries, especially in parts of East Asia and Southeast Asia, the possession of marijuana can carry a long jail sentence and worse, may lead to life imprisonment or even execution.

The Potential Rise of PotCoin

With the continued push for the legalization of marijuana globally, it is almost certain that PotCoin will rise and be a standard form of payment in the multi-billion cannabis industry. Several cryptocurrency experts forecast the price of PotCoin to hit the $0.449 mark by the end of 2018. Despite its seemingly fast growth in the past few months attributed to promotions through celebrity endorsement and charity activities, it is predicted that the price of PotCoin will be $0.42 in five years. This is mainly because other cryptocurrencies are on a high, coupled with the huge competition faced in the cannabis cryptocurrency space.

PotCoin faces enormous competition from DopeCoin and CannaCoin which are also aiming at controlling the marijuana market.

Cannacoin had emerged as a serious competitor with the development team producing a slick new website which expanded the coin's presence to a second crypto exchange, Bittrex, giving its potential adopters two places where they could purchase the coin. However, the coin was criticized for limiting the selection of goods users could buy, despite constant assurances put forward by the developers that they were in negations with the merchants in Washington State Market to accept Cannacoin as a means of payment. Cannacoin has failed to gain credibility and market share, therefore, reducing its competitive ability as a standard marijuana cryptocurrency.

Additionally, the price of Cannacoin has continued to fall, is down to $0.041893, with the market cap being $196,973 and trading volume $148.522 at the time of writing.

On the other hand, PotCoin has been making serious, determined moves in an attempt to control the cannabis banking industry, making almost daily announcements about real, actual developments aimed at assuring their users' continued growth. PotCoin has also redesigned their website (even snagging the .com suffix), keeping their classic silver-white-green motif, yet going for a slick, 21st Century design that is undeniable professional and attractive.

Additionally, PotCoin has more than 100 merchants accepting PotCoin as a means of payment in over 800 locations in the world. PotCoin has also installed and integrated ATMs in over 35 different countries making it the world's second-biggest cryptocurrency ATM provider after Bitcoin. This makes it easier for dispensaries and cannabis vendors to accept PotCoin, making PotCoin transactions easier and simplified.

The price of PotCoin has seen a more-or-less steady upward trend. The price fluctuations are attributed to the once selfish miners who dumped their holdings in exchange for other cryptocurrencies. Nevertheless, PotCoin, by comparison, has by far the biggest market cap, trading volume, and community of users. At the time of writing, the price of PotCoin was USD 0.071008, with a market cap of $15,631,748 and a circulating supply of 220, 140, and 346.

With a consensus algorithm of proof-of-stake (POSV), PotCoin allows users to gain up to 5% interest per year on their stakes which are a very good investment for users.

Nevertheless, past performance is never an assurance of future trends, however, but we have reason to suspect that PotCoin is about to break out in a major way, and that reason was unveiled when PotCoin debuted their pre-paid PotCoin cards, a game-changing advance in cryptocurrency distribution. The largest stumbling block in the consumer adoption of cryptocurrency is access, and preloaded cards will destroy that barrier while simultaneously spurring merchant adoption, creating a positive feedback cycle that will drive the price of PotCoin even higher.

Chapter Six: The PotCoin Cryptocurrency Movement

To promote its brand, PotCoin has continually partnered with celebrities in an aim to gather a huge following. In addition to fundraising, PotCoin has also taken part in some fundraising as discussed below.

Charitable Fundraising

Snoop Youth Football League

On February 16, 2014, the PotCoin community paid $500 to the Snoop Youth Football League; a charity set up by Snoop Dogg. The Snoop Youth Football League is a non-profit organization that gives inner-city children between the ages of 5 and 13 the chance to participate in youth football and cheer. It took the community two days to raise the required 55,000 PotCoins. A video message was uploaded by Snoop Dogg, thanking Nick Iverson, a PotCoin developer for the contribution, and prizes were sent to some community members.

Cannabis Health Service

On June 6, 2014, 18,500 PotCoins were paid to the Cannabis Health Service, an organization based in the United Kingdom, after a community on PotFunder. The Cannabis Health Service was founded by Colin Davies, a prominent activist with the aim of helping patients who believe, or have found, cannabis to be beneficial for the treatment of their illnesses.

FRAXA Research Foundation

On July 8, 2014, a check for US$2,000 was presented to the FRAXA Research Foundation after a community fundraising initiative through PotFunder. The fundraising campaign was pioneered by PotCoin team member Russell Thomas who has a son that suffers from Fragile X syndrome (a form of mental retardation and the known leading cause of autism). FRAXA is a non-profit organization, their mission is to find effective treatments and ultimately a cure for Fragile X.

Healthy Hopes

On August 2, 2014, 32,578 PotCoins were paid to Healthy Hopes after a campaign on PotFunder. Healthy Hopes are a charity whose mission is to give safe access to medical cannabis to chronically and seriously ill patients who have no hope of legally obtaining the medicine they need where they live.

Celebrity Endorsement

Celebrity endorsement has become one of the most commonly used marketing tactics in the crypto space. In an industry where fringe altcoins are finding it increasingly difficult to come up with unique and fresh narratives, associating the brand with a celebrity's name has emerged as a tempting and potentially lucrative alternative. Cryptocurrencies like Centra and Stox have already established the blueprint for celebrity endorsement by hiring the likes of DJ Khaled, Floyd Mayweather, Liang Tian, and Luis Suarez to promote their brand. PotCoin has also endorsed Dennis Rodman as its ambassador.

Dennis Rodman's Trip to the North Korean Summit

Dennis Rodman, an ex-NBA player hall of fame, has been endorsed by PotCoin which sponsored him to the North Korea summit in Singapore where US President Donald Trump and North Korean leader Kim Jong-un historically met. This was the second time Rodman has traveled to North Korea with the sponsorship of PotCoin. The digital coin stepped in as Rodman's sponsor at the beginning of 2015, saying, "We believe in Dennis Rodman's mission to bring peace to the world." The cryptocurrency backed his 2017 trip to North Korea, and as part of the arrangement, Rodman wore PotCoin merchandise, including a T-shirt and hat.

Rodman has flown to Pyongyang five times since Kim Jong-un became North Korea's leader. Previously, his trips were sponsored by Irish betting website Paddy Power. It eventually severed the relationship in December 2013, citing "The worldwide scrutiny and condemnation of the North Korean regime in recent weeks."

Rodman's trip to the North Korean summit hit several headlines stating that PotCoin prices were "soaring" as a result of Rodman's endorsement, but the coin isn't worth much, to begin with. A double-digit increase means it has only gone up a couple of cents in value. The market response to the cleverly publicized Rodman's trip to the North Korean summit was almost immediate. After 24 hours, the price of PotCoin had increased by more than 18%, reaching the 10 cents per coin mark for only the second time in the coin's history. The overall market cap of the digital currency surpassed $20 million and the trading volumes for that day were just below $1 million.

Chapter Seven: The Future of PotCoin and Block Chain Technology

The Future of Blockchain

Blockchain refers a distributed ledger technology providing a platform for the operations of common cryptocurrencies such as Ethereum and Bitcoin. In a blockchain, data can be transferred and recorded in a transparent, auditable, safe and resistant to alterations kind of way. This technology can make the organizations that use it open, democratic, decentralized, efficient, and secure.

What is Changing?

- Blockchain will be adopted by central

banks, and cryptographically-secured currencies will become widely used.

- Nasdaq will launch blockchain-enabled digital ledger technology that will be used to expand and enhance the equity management capabilities offered by its Nasdaq Private Market platform.

- The settlement of currency, equity, and fixed income trades almost instantaneously through permissioned distributed ledgers and develops an efficient way in which financial institutions can manage asset and finance.

Control

- Blockchain technology will offer

identity authentication through an openly visible ledger thus significantly reducing cybercrime.

- Car rental agencies could use smart contracts that automatically allow rentals when payments are received, and insurance information authenticated through a blockchain.

- Sensors placed on refrigerators which in turn integrated to a blockchain via the internet could effectively automate interactions with users - anything from ordering and paying for food, to arranging for its own software upgrades and tracking its warranty.

- Small businesses could use blockchain to create trusted trading platforms among themselves.

- Blockchain could potentially help bring robustness and transparency to the post-trade environment.

- New technologies such as blockchain have the potential to reduce cyber risks by offering identity authentication through a visible ledger.

- A bank could pay the supplier instantly over the Internet.

- Blockchain technology will alter timing on risk.

Crime

- A new blockchain start-up has claimed its software could help track down criminals faster and cheaper than ever.

- Connecticut is warning parents that a new Darknet cryptocurrency called Bitcoin could be to blame for helping underage drinkers to get buzzed.

Industries Likely to be Disrupted by Block Chain Technology

With blockchain technology having a clear future, it is certain that most industries in the coming 5 to 10 years will be disrupted by this technology. Here are some of the industries:

1. Banking and Payments

Blockchain developers are predicting that the technology will significantly change the banking and finance sector in a way similar to social media changing the media industry. Many people worldwide will be able to access financial services through blockchain technology. For instance, Bitcoin allows anyone to send money across borders almost instantly and with relatively low fees.

Many banks like Barclays are also working on adopting blockchain technology to make their business operations faster, more efficient and secure. Banks are also increasingly investing in blockchain start-ups and projects.

2. Cyber Security

Although the ledger for blockchain is public, the data is verified and encrypted using an advanced type of cryptography. The data is now less prone to being hacked.

That said, the applications built on the blockchain are still young, and there have been several hacks in recent months. This is something future applications will need to grow up to.

3. Supply Chain Management

With blockchain technology, transactions can be documented in permanent decentralized records in addition to being overseen securely and openly. This will significantly reduce delays and errors associated with humans. Additionally, the supply chain of products will be efficiently monitored for cost, labor, wastage and even emissions in the line of production. Blockchain technology can be used to ensure that only verified authentic products are distributed in the market.

4. Forecasting

The blockchain technology will change research methods including consultation, analysis, and forecasting. Market research will be done through online platforms such as Augur. Additionally, predicting bets from sports or stock markets to elections could easily be monitored

5. Networking and the Internet of Things (IoT)

Samsung and IBM are already using blockchain to develop a new technology called ADEPT used to link IoT devices to a decentralized network. This technology will be a change in the communication industry as it will eliminate the need for central control in device communication. Devices under ADEPT will carry out communication among itself thus automatically updating software, control bugs and manage the rate of energy use.

6. Insurance

Since the insurance industry is based on trust, blockchain technology can improve trust in insurance contracts by verifying and authenticating data issued when writing an insurance contract. For instance, Oracles can be used to integrate real-life data into smart contracts thereby significantly improving insurance policies relying on real-life data for instance crop insurance.

7. Private Transport and Ride Sharing

The blockchain can be used to make decentralized versions of peer-to-peer ridesharing apps, this allows both car owners and users to arrange the terms and conditions in a secure way without the need of third-party providers. Start-ups working in this area include Arcade City and La'Zooz.

The use of built-in e-wallets can allow car owners to automatically pay for parking, highway tolls, and electricity top-ups for their vehicle. UBS, ZF, and Innogy are some of the companies developing blockchain-based e-wallets.

8. Cloud Storage

Blockchain technology will improve data security on cloud storage which is usually exposed to hacking, loss and human error.

9. Charity

Charity donations are usually associated with theft, economical use, and corruption. Through the use of blockchain technology, charity donations will be tracked to ensure it ends up performing the purpose it was meant for. For instance, Bitcoin-based charities such as BitGive Foundation use blockchain technology to secure donated funds.

10. Voting

Voting around the globe is usually faced with claims of rigging through voter bribing or double registration. This could soon end with blockchain technology being used for voter registration, identification and electronic vote counting. With blockchain, no data can be changed therefore no cast votes can be changed or removed making elections more democratic and fair. Democracy Earth and Follow My Vote are two start-ups aiming to disrupt democracy itself through creating blockchain-based online voting systems for governments.

Future of PotCoin and other Marijuana-Based Cryptocurrencies

The future of PotCoin is moderately bright, and you should certainly invest in it. The latest investing craze combines two things that are hot right now: Legal marijuana and cryptocurrencies. But whether pot-based cryptocurrencies are a match made on high, or somewhere else, may depend on who you talk with. PotCoin may have a better chance of being a standard marijuana coin thanks to Dennis Rodman's endorsement which sparked public debate, in turn, marketing the coin globally.

There are now at least five well-known companies in the marijuana cryptocurrency market, start-ups which were developed and launched with marijuana being legalized in most parts of the world including twenty-nine states in the USA. However, marijuana being a high-risk industry, its legalization is still in doubt attributed to the strict anti-marijuana Trump government.

Nevertheless, even with the legalization of medical marijuana, business dealing in medical marijuana have continually struggled to build a financial relationship with the banking industry, with many financial institutions not offering banking services customized for the marijuana industry.

Some pot-based businesses in California have managed to open bank accounts, but they are not rushing to advertise it, notes Sahar. PotCoin has created an opening in the marijuana industry which when correctly exploited can provide a lasting banking solution to business dealing in marijuana in addition to eliminating cash-only transactions which could be dangerous.

PotCoin launched in 2014, was developed with a mission to provide banking solution in the cannabis industry. In a recent report, the company said that they had installed more than 800 PotCoin ATMs in over 35 countries to enable seamless transactions among its users.

With a market cap of $33.8 million from

$81,547 in 2014, PotCoin is undoubtedly the most common cannabis cryptocurrency toping Cannabiscoin with a market cap of 6.5 and DopeCoin with a market cap of $9.5 million.

CannabisCoin targets the medical dispensary sector, but with a twist: Each CannabisCoin can be exchanged with 1 gram of medical pot. DopeCoin appears focused on product development, having just released the "beta version" of "Crypto Billings," which it boasts will "allow retailers to quickly adapt our secure cryptocurrency payment method into their online retail operations."

Comparing HempCoin with a market cap $41 million and CannaCoin with a market cap of $413,769, both are struggling to find followers in the cannabis sector. While to the pot industry's banking challenges, dispensaries and other pot establishments have been slow to embrace the proffered solution.

There is a need for a financial solution for the growing industry of marijuana, and that filling that niche could indeed be profitable. "The appeal is figuring out a viable solution to the cannabis cash problem - it is a very lucrative area, and a large part of my career is built on cannabis banking," she said. Pot dispensaries and retailers still do need conventional banking services and can't just fly off into the marijuana cryptocurrency sunset and write off all financial institutions, Ayinehsazian says.

Yet, banks are already highly nervous about accepting marijuana cash by the fact of illegality associated with marijuana. Additionally, the introduction of cannabis cryptocurrencies will not help the situation as banks will still be unwilling to trade funds from cannabis sales.

With marijuana-based cryptocurrencies future faced with uncertainty, many investors see the light at the end of the tunnel. One attraction to investors is the Bitcoin factor, with the once-obscure digital currency has seen its value soar from a couple hundred dollars a little more than a year ago, to just under $8,800 as of February 13.

Investors in the cannabis industry are compelled to invest in the cryptocurrencies with a hope that marijuana will be declared legal worldwide to gain huge returns on their investments.

In the near future, there is a substantial potential in marijuana being traded as an agricultural product similar to how corn and other agricultural products are being exported and marketed worldwide. Marijuana trade may just be one of the biggest future markets in the world generating billions in trade profit, and thus, marijuana-based cryptocurrencies could be having a bright future worth investing in.

Conclusion

Should I Invest In PotCoin?

With marijuana legalization from the beginning of 2004 coupled with the opening of numerous marijuana stores in several parts of USA notably, Colorado, the launch of PotCoin a marijuana-based cryptocurrency was greeted with excitement. The main aim of PotCoin was to enable cannabis enthusiasts to buy their favorite cannabis products anonymously without the need of having hard cash, in turn, receiving great offers on using the coin as a means of payment.

With the filled excitement greeted with its launch, PotCoin suddenly fell into recesses after few months after heating news headline failing to get up a serious following. The currency was taken over by a new set of administrators, including the developer of a PotWallet/PayPal-esque service created to make using PotCoin even more practical.

Hope springs eternal, and now, more than two years after the vaporware debut of PotCoin, there may be the significant hopes in reviving the currency with PotCoin developers completely overhauling the currency couple with community efforts to keep the coin alive.

Among other internal changes, the company acquired its own data center in Montreal in an effort to own their hardware and network to improve the coin's scalability, reliability, and security.

Chatter in online forums, like its own and Reddit, sedates, as is activity on Facebook and among its 437 Instagram followers.

With only 12 business listed as accepting PotCoin as a means of payment, the coin is undoubtedly still under development. However, PotCoin has some promising features that could easily attract investors. For instance, transactions on PotCoin and PotWallet is free and just within 40 seconds compared to Bitcoin's 10 minutes long wait. Fast transactions would be imperative for a customer-merchant payment service thereby enabling seamless interaction between buyers and sales.

Cannabis consumers could easily be attracted to invest in PotCoin to as they identify themselves with the coin in addition to generating interest on their initial investment value. With the currency being highly volatile, investors should critically analyze their investment before placing their hard-earned money in

PotCoin investment.

Some people dispute the need for a standard cryptocurrency in the cannabis space to solve the problem of banking in the industry. They argue that with marijuana legalization by the federal government in Canada and the United States, banks will start accepting proceeds from the cannabis industry erasing the banking problem which forms the basis of PotCoin development. However, at the moment, banks don't seem to be developing financial relationships with the cannabis sector as with the situation with two of Canada's biggest lenders shunning all cannabis-related business despite its legalization.

There are all these advantages for the cannabis industry because they get hit with a lot of fraud and get stuck with a chargeback, in which dispensaries owe credit card companies for fraudulent charges. While most dispensaries still can't accept major credit cards, apparently that's a credit card company's way of thanking dispensaries for their business. If dispensaries don't like it, they can ditch the plastic, or the heaps of cash, for PotCoin, he said. "That's the beauty of it."

Cryptocurrency experts caution against getting too feverish, at least until PotCoin gains many, many more users, beginning with traders, or rather business accepting PotCoin as a means of payment. "I'm a little leery," the business professor said. "They need to get more merchants on board. If they do that, the people will follow, because why would I want PotCoins if I can't use them?"

Cryptocurrency aficionados may take a

large interest in another huge change that has been made to PotCoin: It's no longer minable. It now runs on a proof-of-stake system where users earn 5% to 7% interest on their currency. To cut it, mining is the process that makes new units of certain cryptocurrencies, including Bitcoin, but it is frowned upon by many as it puts a strain on the computer hardware.

PotCoin's description on their Facebook page says it's "Aspiring to become the standard form of payment for the legalized cannabis industry." Therefore, on investing in PotCoins, I would advise you to wait until it gains lots of followers or marijuana is fully legalized. Nevertheless, if you use marijuana for both medical and or recreational use, you should certainly invest in PotCoin as there are numerous benefits, such as a discount on cannabis products such as crypto juice and cannabis seeds.

In summary, PotCoin is looking forward to be a standard form of payment in the cannabis space by creating customer's loyalty in using the coin. Additionally, the development team seeks to extend stability, security and credibility in the growing marijuana community. In doing this, PotCoin has a number of programs such as the merchant program to attract new companies to accept PotCoin as the form of payment. The marketing team is also developing features meant to enhance seamless transactions. For instance, the "Pay it Forward" feature through Potfounder.com allows the use of PotCoin in crowd-funding thus small projects can easily be funded through this feature. PotCoin has also been very active in charity taking part in a number of charity activity notably, "Support Cannabis for Fragile X Syndrome" which is meant to support people with Fragile X Syndrome known for causing autism and intellectual disability especially among boys.

PotCoin has also some activities in which marijuana enthusiasts can engage take part i.e. the developer program and ambassador program. The role of ambassadors is to market PotCoin by spreading the word about the coin to business while developers are tasked to come up with innovations meant to improve the functionality of the coin. Will PotCoin eventually develop to the standards of Bitcoin. Time will tell!

I hope this book has very been informative on PotCoin, which is aiming to become a standard form of payment for the legalized cannabis industry. Thank you!